TREES AND SHRUBS

valuable to bees

by M. F. Mountain

Bee Research Association

Jointly published by: The International Bee Research Association,
a Company Limited by Guarantee, 1, Agincourt Street, Monmouth, NP25 3DZ
(UK) &

Northern Bee Books, Scout Bottom Farm, Mytholmroyd,
Hebden Bridge HX7 SJS (UK).

Obtainable from:
www.ibra.org.uk & www.northernbeebooks.co.uk

© 2021 The International Bee Research Association.

ISBN: 978-1-913811-08-2

Cover design and artwork by DM Design & Print
Cover image - Stuart A. Roberts - 'Honey bee on privet'
Rear cover image - Stuart A. Roberts - The ivy bee, *Colletes hederae*,
foraging on ivy.

TREES AND SHRUBS VALUABLE TO BEES

by M. F. Mountain
Domestic Horticultural Officer to Buckinghamshire County Council

PREFACE

This list is a sequel to the Association's leaflet "Save our Pollinating Insects". We have received many enquiries as to how those with opportunities to plant trees and shrubs on public or private land can implement the plea contained in the leaflet - to conserve our economically useful pollinators.

The books and articles available have proved curiously inadequate in assessing the value to bees of trees and shrubs that can be recommended for planting on their horticultural merit. The best source of information has been a leaflet by Miss M. F. Mountain, published in 1950 by West Sussex County Beekeeping Advisory Service and the Chichester Division of the Sussex Beekeepers' Association. The present leaflet represents a complete revision of this earlier one, with many additions; it has been compiled on the basis of fifteen further years of observation, and the published and unpublished records of the Bee Research Association.

May 1965

Eva Crane
Director, Bee Research Association

GENERAL NOTES ON THE TABLES

1. When planning amenity plantings for roads, housing developments, factories, schools and other public buildings, the best results are likely to be obtained if shrubs are grouped with young trees when these are planted, in order to give them protection (especially from wind) during their first few years. In parks, and in public and private gardens, more shelter is commonly available from existing features.

2. In the third column of the lists:

N   indicates a source of nectar

P   indicates a source of pollen

-   in place of either N or P indicates no supply in normal conditions in Britain

?   indicates that sufficient knowledge is not yet available about the plant in Britain, and that personal observations are invited by the Bee Research Association. Many plants will grow and flower in a wider variety of conditions than those which enable them to secrete nectar.

3. Plants normally growing with a trunk are included in the tree list (pages 12-14). All others are in the shrub list (pages 5-10) except for the climbers (page 11), which are not freestanding.

4. The season listed is for southern England in a "normal" year. A great variation can occur from year to year, especially in early spring flowering dates; see for instance "A calendar of bee plants" by Dorothy Hodges (page 15).

5. The following abbreviations are used:
   dec.   deciduous
   ev.   evergreen
   fl.   flower, flowering
   fls.   flowers
   fr.   fruit
   frs.   fruits
   sp.   species
   spp.   species (plural)
   vars. varieties

This list includes bushy subjects (not normally grown on a trunk) suitable for use in gardens, on road islands – anti-dazzle belts and roundabouts – and on either flat or sloping verges. Species have been selected because of their attractive flowers or fruit, foliage or stem colour. A few less showy bushes of outstanding value to bees are also included.

Botanical name	Common name	Value to bees	Height (ft.)	Spread (ft.)	Flowering season	Notes
Amelanchier canadensis	snowy mespilus	- P	20-30	20-30	April	flowers, autumn foliage
Aralia elata (chinensis)	Chinese angelica	N ?	to 30		Aug-Sep	large handsome foliage
Arbutus unedo	strawberry tree	N P	15-25	18-25	Oct-Nov	if mild, frs.&fls.together
Berberis all spp., esp.:	barberry					not in U.S.A. because of [black rust
B. darwinii		N P	8-10	8-10	Apr-May	ev., deep gold flowers
B. irwinii		N P	2	6-8	Apr-May	red berries
B. prattii		N P	6-8	6-8	Apr-May	ev., narrow lves.
B. stenophylla		N P	10-12	10-15	Apr-May	autumn colour
B. thunbergii & var.atropurpurea		N P	6-8	7-9	Apr-May	masses of red berries; dec.
B. wilsoniae		N P	3-4	6-8	July	
Buddleia globosa		N ?	12-15	10-15	May-June	fls. are golden balls
Buxus sempervirens	box	N P	15-18	15-18	April	sun or shade;stands clipping
Calluna vars.	ling	N P	1-2	2-3	Aug-Sep	acid soil
Caragana arborescens	pea tree	N P	15-25	15-25	May	not very hardy
Ceanothus (spring fl.spp.)		N P	20-25	18-25	March	needs wall shelter
Chaenomeles speciosa = Cydonia japonica: single vars.	Japanese quince	N P	3-4 to 10-12	3-4 to 10-12	Mar-Apr	early pink flowers
Cistus spp.	rock roses	N P	2-2½ to 6-8	2½-3 to 7-9	May-July	not long lived; single flowers
Colutea arborescens	bladder senna	N P	4-5	4-5	June-Sep	yellow flowers, swollen pods

Botanical name	Common name	Value to bees	Height (ft.)	Spread (ft.)	Flowering season	Notes
Cornus alba	dogwood	N P	8-10	9-12	May-June	usually for red winter stems; then pruned hard and no fls.
C. mas	cornelian cherry	N P	15-20	18-25	Feb-Mar	small yellow flowers
C. stolonifera etc.	dogwood	N ?	6-8	7-10	May-June	coloured stems as above
Cotinus coggygria =						
Rhus cotinus	smoke tree	? ?	12-16	12-15	June-July }	flower stalks & autumn
C.c.foliis purpureis		? ?	10-12	12-15		colour; flowers minute
Cotoneaster spp., especially						
C. conspicua		N P	3-4	9-10	June	
C. cornubia		N P	15-20	15++		
C. dammeri		N P	1-2 in.	spreads		
C. franchetii		N P	8-12	8-12	July	berries; some ev., some dec.
C. frigida		N P	15-25	20-40	June-July	many flower in June gap
C. horizontalis		N P	$2\frac{1}{2}$-4	8-10	May	susceptible to fireblight
C. microphylla		N P	$2\frac{1}{2}$-4	8-10	May-June	
C. salicifolia rugosa		N P	15-20	15-20	June	
C. simonsii		N P	10-12	9-10	June	
Cytisus spp.	brooms	N P	$\frac{1}{2}$ to 6-7	$\frac{1}{2}$-1 to 6-8	May-June	flowers yellow and pink shades
Daphne cneorum		N P	1-1$\frac{1}{2}$	3-4	Apr-May }	scent; not long-lived
D. mezereum	mezereon	N P	3$\frac{1}{2}$-4$\frac{1}{2}$	3$\frac{1}{2}$-4	Feb-Mar }	
Deutzia spp. & vars.						
D. kalmiflora		? ?	4-5	5-8	May-June	very pale pink fls.
D. scabra		? P	8-12	6-8	June-July	white flowers
Diervilla spp.		N ?	6-8	6-8	May-June	showy flowers

Elaeagnus spp. e.g.		? ?	9-12	10-15	Oct-Nov	grown for foliage; ev.
E. pungens variegata						variegated forms
Erica spp. including	winter heather					tolerate alkaline soil
E. carnea		N P	ca.1	ca.1½	Nov-Feb	
E. mediterranea		N ?	5-7	6-8	Feb-Mar }	acid soil only
E. tetralix		N P	½-1	1½-2	June-Oct }	
E. vagans		? ?	1½-2	2½-3½	Sep-Nov	
Escallonia spp. & hybrids e.g.						ev.; pink flowers
Donard seedling		N ?	9-12	9-12	June	
Euonymus alatus	winged spindle tree	? ?	6-8	6-10		autumn colour
Fuchsia		N P	4-8	4-8	June-Oct	where hardy, sometimes herbaceous
Genista						
Good ones are:						
G. aethnensis	Mount Etna broom	? ?	15-20	15-18	July	pale yellow
G. hispanica	Spanish gorse	? ?	2½-4	5-8	May-June }	yellow mounds in season, then green
G. lydia		? ?	2½-3	6-8	May-June }	
Hamamelis japonica	witch hazels	? ?	12-18	12-15	Dec-Feb }	winter flowers
H. mollis		? ?	12-18	12-15	Dec-Feb }	
Hebe pagei		? ?	ca.1	2½-3	May-June	grey foliage; ground cover
Hippophae rhamnoides	sea buckthorn	? ?	15-25	15-20	April	male & female plants; fls. inconspicuous; orange berries
Hydrangea: only fertile flowers		N P	6-8	8-10	Aug-Sep	some spp. attractive; large sterile fls. surround inconspicuous fertile ones
Hypericum androsaemum	St. John's wort	? P	3½-4½	5-6	June-Aug]	yellow flowers
H. calycinum		? P	1-1½	spreads	June-Aug]	
H. patulum		? P	3-4	4-5	June-Aug]	

8

Botanical name	Common name	Value to bees	Height (ft.)	Spread (ft.)	Flowering season	Notes
Ilex aquifolium	holly	N P	40-50	25-35	May	some variegated forms; no fls. when clipped
Kolkwitzia amabilis		? ?	6-8	6-9	May-June	pink flowers
Laurus nobilis	bay	N ?	20-45	15-30	May-June	culinary value
Lavandula species & fertile vars.	lavender	N P	ca.3	2-3	July-Aug	
Lonicera purpusii	winter-flowering honeysuckle	N P	7-9	7-10	Jan-Feb	white scented flowers
L. standishii		N P	7-9	7-10	Jan-Feb	white flowers
Magnolia denudata		? ?	25-30	25-30	May	
M. grandiflora		? ?	30-40	25-35	July-Sep	ev.;scented fls. best by wall
M. kobus		? ?	25-30	25-30	April	tolerates lime
M. soulangiana & vars.		? ?	20-30	30-40	Apr-May	white & pink fls., purple forms
M. stellata		? P	10-15	12-20	Mar-Apr	starry white flowers
Mahonia aquifolium		N P	3-5	3-5	Feb-May	good ground cover; very hardy; yellow flowers
M. bealei		N P	5-7	8-15	Feb-Mar	handsome foliage;yellow scented flowers
Mespilus germanica	medlar	N P	20-25	20-25	May-June	outstanding white fls.,fr. edible
Olearia haastii	daisy bush	N P	6-9	8-12	July-Aug	hardy; white daisy flowers
Osmanthus delavayi		? ?	7-10	9-12	April	scented white fls.; ev.
O. ilicifolius		? ?	7-9	8-10	Sep-Oct	ev.; tolerates shade
Perovskia atriplicifolia		N ?	6-8	7-9	Aug-Sep	grey upright stems; blue fls. not long-lived
Philadelphus vars.	mock orange	? ?	$3\frac{1}{2}-4\frac{1}{2}$ to9-12	$3\frac{1}{2}-4\frac{1}{2}$ to9-12	June-July	white scented flowers

Species	Common name	N P			Flowering	Notes
Photinia villosa		? ?	12-18	12-18	May	grown for berries & odd red leaves in autumn
Potentilla spp. e.g. fruticosa		N P	3½-4	5-7	June-Sep	yellow/white flowers
& its vars. e.g. arbuscula		? ?	3½-4	8-9	May-Oct	very long fl. period
Prunus laurocerasus	cherry laurel	N P	15-20	20-30	April	extrafloral nectar also at other times
P. lusitanica	Portugal laurel	N P	10-15	15-20	June	ev.; large leaves
Pyracantha vars.	firethorn	N P	14-16	14-16	May-June	ev.; berries tolerate N. walls; susceptible to fireblight
Rhamnus cathartica	common buckthorn	N P	6-8	6-8	May	very small green fls.; black berries; chalk
R. frangula	alder buckthorn	N P	6-8	6-8	May	
Rhus typhina	stag's horn sumach	N P	10-15	12-16	July	autumn foliage
R. glabra		N P	4-6	4-6	July-Aug	
Ribes sanguineum	flowering currant	N P	8-10	9-12	Mar-Apr	pink flowers
R. speciosum		N P	8-10	9-10	May-June	red flowers
Rosa single spp.		? P	bush & climbers		May-Sep	dog rose; many single spp.
R. rugosa	Ramanus rose	? ?	5-8	5-8	June-July	
Rosmarinus officinalis	rosemary	N P	6-7	6-7	Apr-May	blue flowers; ev.
Rubus deliciosus		? ?	6-10	6-10	May-June	large pink or white fls.
R. odoratus		? ?	6-8	6-8	July-Sep	
Salix caprea	goat willow	N P	15-20	18-25	Feb-Mar	earliest; male plants; showy
S. medemii		N P	12-14	12-14	Feb-Mar	
S. repens	creeping willow	N P	4-7	8-12	April	quite flat on rocks
S. smithiana		? ?	15		Mar-Apr	
Senecio greyi		? P	4-5	6-8	June-July	grey foliage
Skimmia japonica		N P	1½-2	2-2½	March	ev. berries on plants

10

Botanical name	Common name	Value to bees	Height (ft.)	Spread (ft.)	Flowering Season	Notes
Symphoricarpos spp.	snowberry	N P	5-6 to 6-8	5-6 to 9-12	June-July	many spread (by suckers)
Syringa spp. & vars.	lilac	? P	15-18	15-18	May	
Tamarix pentandra	tamarisk	N P	12-15	15-18	July-Aug	pink fl. heads; fls. very small
Ulex europaeus	gorse	N P	4-5	5-7	Apr-May	fls. other times also
U. nanus	dwarf gorse	N P	$\frac{1}{2}$-1	3-4	Sept	
Viburnum bodnantense		? ?	9-12	10-12	Dec-Feb	pale pink, scented
V. burkwoodii		? ?	8-10	9-12	April	half ev.; scented
V. davidi		? ?	2-3	4-5	June	ev.; grown for leaves
V. opulus	guelder rose	N ?	10-15	12-18	May-June	white
V. tinus	laurustinus	N P	7-10	8-10	Jan-April	ev.; buds pink, fls. white
V. tomentosum		? ?	9-12	10-15	May	fls. white
Weigela florida & hybrids		N ?	6-8	6-8	May-June	showy pink, white & red fls.

Botanical name	Common name	Value to bees	Flowering season	Notes
Parthenocissus tricuspidata.	Virginia creeper "ampelopsis"	N P	August	handsome fruits
Celastrus		? ?	June etc.	white & pale pink scented fls.
Clematis armandii		? P	April-May	white or pink fls. not scented; suits alkaline soil
C. montana		N P	April-May	
Eccremocarpus		? ?	July on	often behaves as annual; red tubular flowers
Hedera helix	ivy	N P	Sep-Dec	self-clinging; stands some shade
Hydrangea petiolaris		? ?	June	
Schizophragma integrifolia		? ?	July	
Vitis spp.	vine	N P	June etc.	nectar freely only in warm places
Wistaria sinensis	wistaria	N P	May(Aug.)	nectar only in warm weather

TREES

The sizes given here are for mature specimens and will not be attained for at least thirty years; if conditions are unfavourable, full stature will never be attained. Some trees suitable for amenity planting are too big for any but the largest gardens. Although they may grow slowly, those whose listed height is 50 feet or more may need to be lopped, and thus disfigured, before they reach maturity. It is in general unwise to plant any tree listed at 30 feet or more in a town garden of a quarter of an acre or less.

Botanical name	Common name	Value to bees	Height (ft.)	Spread (ft.)	Flowering season	Notes
Acer campestre	field maple	N P	15-40	15-20		
A.ginnala		? ?	15-20	15-20		
A.griseum	paper bark maple	? ?	20-40	15-20		fls. not showy; some spp. of value for bark, others for foliage
A.negundo vars.	box elder (U.S.A.)	)N P	40-60	35-40	Apr-May	
A.palmatum & vars.	Japanese maples	? ?	10-14	15-20		
A.platanoides & vars. eg. Goldsworth purple	Norway maples	N P	60-70	40-50		
A.pseudoplatanus	sycamore	N P	70-80	50-60		
Aesculus hippocastanum	horse chestnut	N P	80-100	50-70	Apr-May	
A.carnea		N P	30-50	25-40	2 wks. later	crimson fls.
A.indica		N P	80-100	50-70	3-4 wks. later	flowers in "June gap"
A.pavia	red buckeye (U.S.A.)	N P	12-20	12-20	June	red flowers
Ailanthus altissima	tree of heaven	N P	60-90	40-50	July-Aug	tolerates city atmosphere

Alnus cordata	Italian alder	?	?	60-80	30-40	March	very good in wet places
A.glutinosa	common alder	-	P	50-70	25-35	Feb-Mar	
A.incana	grey alder	-	P	60-70	30-50	Feb	
Betula spp.	birches	-	P	20-60	20-30	May	some spp. very beautiful bark; graceful
Castanea sativa	Spanish chestnut	N	P	50-60	50-60	July	flowers inconspicuous
Catalpa bignonioides	Indian bean	N	P	30-50	30-50	July-Aug	showy white fls.;extrafloral nectaries
Cercis siliquastrum	Judas tree	N	P	12-15	12-15	May-June	pretty pink pea flowers
Crataegus: single vars. of oxyacantha & monogyna	hawthorn	N	P	20-30	25-35	May	most give fls. & berries
C.crus-galli	cockspur thorn	?	?	15-18	18-20	June	susceptible to fireblight
C.prunifolia & others		N	P	15-18	20-25	June	C.crus-galli gives autumn colour
Davidia involucrata	handkerchief tree	?	?	40-60	30-40	May	large white bracts
Fagus sylvatica vars.	beech	-	P	80-120	70-90	Apr-May	short fl. season
Fraxinus spp.	ashes	-	P	50-70	30-50	May	
Ilex aquifolium & vars.	holly	N	P	40-50	25-35	May	some with variegated foliage
Koelreuteria paniculata		N	?	10-40	20-30	July-Aug	
Liquidambar styraciflua	sweet gum of N.America	N	?	50-70	40-50	spring	autumn colours
Liriodendron tulipifera	tulip tree	N	P	60-90	35-40	summer	fruits & flowers
Malus in variety	crab apples	N	P	20-40	20-40	Apr-May	not on chalk
Nothofagus procera	southern beech	-	P	to 80			
Populus nigra	black poplar	-	P	to 100		Mar-Apr	
P.tremula	aspen	-	P	to 50		Feb-Mar	

Botanical name	Common name	Value to bees		Height (ft.)	Spread (ft.)	Flowering season	Notes
Prunus amygdalus	almond	N	P	20-25	25-30	Feb-Mar	deep pink flowers
P.avium (single)	wild cherry	N	P	60-70	35-40	Apr	autumn colour, spring fls.
P.cerasifera	myrobalan plum	N	P	25-30	25-30	Feb-Mar	v.pale pink flower
P.padus	bird cherry	N	P	40-60	30-40	May	autumn colour, spring fls.
P.persica (single)	peach	N	P	15-20	18-25	Apr-May	flowers, fruits
P.sargentii		N	P	25-30	18-25	Mar-Apr	autumn foliage; flowers
P.serrulata (single & semi-double vars.)	Japanese cherry	N	P	ca.20	ca.20	Mar-Apr	flowers; alkaline soil
P.spinosa	blackthorn	N	P	12-15	15-18	Mar-May	white flowers
P.subhirtella	spring cherry	N	P	20-25	20-25	April	pendulous; and winter-fl. forms also
P.yedoensis	Yoshino cherry	N	P	25-30	25-30	April	earliest Japanese cherry
Quercus spp.	oaks	-	P	40-80	15-60	May	short flowering season
Robinia pseudoacacia & vars.	false acacia	N	P	60-80	40-50	June	good town tree
Salix alba & vars.	white willow	N	P	70-80	70-80	Apr-May	
S.fragilis	crack willow	N	P	70-80	70-80	Mar-April	winter stems; damp places
S.purpurea e.g. var.pendula							
Sorbus aria	whitebeam	N	P	15-20	25-30	May	white backs to leaves
S.aucuparia	mountain ash	N	P	30-40	30-40	May	red berries early
S.hupehensis		N	P	30-50	25-40	May-June	white or very pale pink
S.intermedia	Swedish whitebeam	?	?	to 40		June	good in towns [berries
Tilia esp. euchlora	Crimea lime	N	?	50-60	30-40	May	less honeydew than some
Ulmus spp.	elms	-	P	40-100	30-50	Feb-Mar	subject to branch dropping & Dutch elm disease

REFERENCES

Bee Research Association (1964) Selected list of publications on the value of different plants as bee forage. Bibliogr. Bee Res. Ass. No. 2

Glukhov, M. M. (1955) [Honey plants] Moscow : State Publishing House for Agricultural Literature 6th ed. [In Russian]

Harwood, A. F. (1947) British bee plants. Foxton : Apis Club

Hodges, D. (1958) A calendar of bee plants. Bee World 39(3) : 63-70

          Flowers for bees month by month. London : Bee Research Association

Howes, F. N. (1945) Plants and beekeeping. London : Faber & Faber

Lovell, H. H. (1926) Honey plants of North America. Medina, Ohio : A. I. Root Co.

Maurizio, A. (1959) Blüte, Nektar, Pollen, Honig. Dtsch. Bienenw. 10

Mountain, M. F. (1950) Shrubs and trees valuable to the honey bee. W. Sussex County Beekeeping Advisory Service and Chichester Division Sussex B. K. A.

Nyárády, A. (1958) A méhlegelo és növényei. Bukarest : Mezógazdasági és Erdészeti Allami Könyvkiadó [In Hungarian]

Pellett, F. C. (1947) American honey plants. New York : Orange Judd Publishing Co. 4th ed.

Pritsch, G. (1959) Sicherung der Bienenweide durch zweckentsprechende Gehölzauswahl in der Landschaft. Dtsch. Landw., Berl. 10(5) : 254-256

www.ingramcontent.com/pod-product-compliance
Lightning Source LLC
LaVergne TN
LVHW070838080426
835511LV00023B/3471